Minnie Rose Lovgreen's RECIPE FOR RAISING CHICKENS

as told by Minnie Rose Lovgreen

edited by Nancy ReKow and Claire Frost

illustrated by Elizabeth Hutchison

hand-lettered by Nancy ReKow

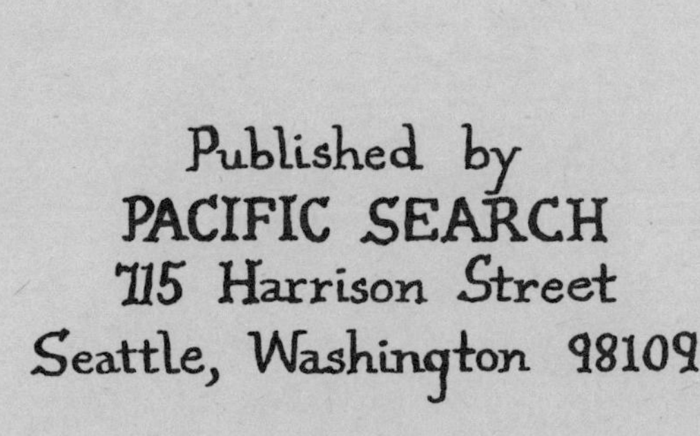
Published by
PACIFIC SEARCH
715 Harrison Street
Seattle, Washington 98109

Second edition, May 1975

First edition printed on Bainbridge Island, Washington, March 1975

International Standard Book Number 0-914718-09-6
Library of Congress Catalog Card No. 75-16891
Printed in the U.S.A.

After sixty years experience raising chickens I wrote this book to help others get started on chickens with the least cost possible. The best of luck to you and your chickens. May you find their ways as interesting as I have.

I want to dedicate my book to my dear friend Nancy Rekow who helped me to put the story together.

Yours sincerely,
Minnie Rose Lovgreen

I specially want to thank all those who helped me so much with the book: Bonnie Boatman, Kern Devin, Ed Doremus, Claire Frost, Gary Frohning, Elizabeth Hutchison, Betsy Leger, Karen Lehne, Nancy ReKow, Kathy Roberts, Caroline Roth, Guenther Roth, Rochelle Schlieps, Lina Tanner, Macy Worden, Page Wightman, and my own family.

"The main thing is to Keep them happy."

TABLE of CONTENTS

I.

THE BROODY HEN AND HER EGGS

In the early spring, about February or March, a bantam hen gets broody. That means she's made up her mind to start a family.[1] She makes a nest somewhere, like in a box or under a log, where she thinks there won't be any rain coming down her back. I once put a piece of plastic over a log, and the hen stuck right with her nest under that. She likes to face southeast or southwest, so she gets morning or afternoon sun. She hardly ever takes the north side. She prefers an outside place where it's clean and there aren't any fleas. Especially if it's early spring, there'll be fleas hatching out in the chicken house from flea eggs laid all winter.

She's very secretive about her nest. She'll come out and eat, and all of a sudden when you think you'll watch her go back, she's suddenly disappeared. You watch another day. You know she must be somewhere in that area, but she's hard to find.

1. Other Kinds of hens can be mothers too, but bantam hens have a natural mothering instinct. That's why I always use the bantams for setting. For more on the bantam, see Section VI., p. 26.

She starts laying an egg each day in her nest. She sits on the nest a little longer every day, a little longer every day, until she lays her last egg. She may lay anywhere from 10 to 14 eggs. She is almost ready to set, but she still has one important job to do. With her beak, she picks some feathers off her breast. Then she scuffles the rest all off, until her breast is bare. She knows she has to set with that bare skin right on her eggs, because there's no heat in feathers. The heat to hatch her eggs comes right from her. She imparts her heat to her babies.

When she stays on her eggs all night, you know she has started setting. Every day she turns all her eggs over with her beak. She only takes about 15 minutes recess off the nest every day, or maybe 30 minutes when the sun shines. She eats and drinks and takes a dust bath, and cleans her beak thoroughly before she goes back on the nest again. For beak cleaning, first she wets her beak, then she rubs it from side to side on the grass. She takes a little more water, swishes it around, then rubs her beak dry on the grass again. She's very particular about having a clean beak to go back on the nest. I don't know whether she thinks the eggs might get messy when she rolls them over. When you see her clean her beak, you know she's just about ready to go back on the nest. She runs like mad then. She only takes that one recess a day. She does that for 21 days.

If you peek in at a setting hen, she says, "Kwark, Kwark, Kwark," in a deep ugly voice, to say, "Don't bother me." She's very protective when she's setting. A person should wear gloves to reach in there because the hen will peck. Her beak is like a hard little fish hook.

The hen talks to her chicks when they're still in the shell, when she hears them beginning to peep, when she hears life coming in the shell. The day before her babies hatch, she doesn't go off the nest at all.

She stays right on it. Say about the 18th day, the baby chick is picking at the shell, trying to make a hole to get out. It peeps, and its mother puts her head under and talks to it, even in the shell. She says, in a deep tone, "Crrrk, crrrk, crrrk," as much as saying, "Hush, everything will be all right." Then the chick pecks away at one side of the larger end of the egg until it makes a little window, and it just keeps that up until it makes a hole big enough to get out. And then it's wet, and the mother sets on it and gets it good and dry with the heat of her body. Then there's more chicks coming out. They all seem to be trying to come out at that same time.

Sometimes a chick finds it difficult to get out of the shell, but from the fact of the hen's body being so warm, it dries up the shell and makes it more brittle, so it's easier to peck a hole and get out. A person should never interfere and try to help the chick out of the shell. The hen does the job much better with her body heat.

A lady once asked me, "Do chicks get milk when they go under their mother?"

And I said, "No."

So she said, "Well, what do they go under there for?"

I said, "They go under there to get their backs warm."

"Well, aren't you afraid she'll mash them when they go under there?"

And I said, "No, she doesn't sit right down on them. She has her legs to stand on, and she just covers them, and she likes to feel them next to her skin where she's taken the feathers off. Have you noticed how they take the feathers off their breast, so their skin is right close to the baby chick? And if you separate the feathers a little bit over, there's a lot of breast space. The hen has a big job to impart her heat to her babies to keep them warm, if they get a little chilled; so she's really a hard-working mother."

"It's really better to take the first-hatched chicks into your house..."

II.

BABY CHICK CARE

I think the first eggs to hatch out are the earliest ones the hen laid. But they all hatch out within a day or two if they're going to hatch.

The hen is very worried when her chicks start to come out of the shell, for fear they'll get hurt. It's really better to take the first-hatched chicks into your house for a while as soon as they've dried off under the hen. Then the hen gives more heat and more attention to the unhatched ones. Whereas if you don't take them in, quite often she'll get excited over some of them and mash them, you see, so you'll find a few dead ones. You're ahead of the game if you take them in the house. Or also, I've seen lively ones hatch out and jump out of the nest, and then the hen doesn't quite know what to do. She can't do anything about it. She can't pick them up by the neck like a cat does, and drag them back. So the chick sits down there and peeps and may die of the cold. Sometimes the hen will go down and sit on it and leave all the other eggs, and that way you lose the whole batch. So I always try to take them in my house.

Of course a hen does manage independently sometimes, if she has to. Maybe you'll look around at a hen and you'll see they're all hatched out, they're all moving, they're all there. The nest is full of chicks and shells, and it's everything combined, and it's wonderful. You say, my goodness, there'll be no taking them into the house. Just leave them all under the hen, and she sits there and cuddles them.

That can happen. But you'll get more chicks in the long run if you always take care of the first ones in your house.

KEEPING CHICKS IN YOUR HOUSE

I put the chicks in a box and pin part of a wool sweater or sock over the box with clothespins, letting it touch their backs, but so they can still get air. The sweater feels like the warm mother sitting on them. Or some people use an electric light. I prefer the sweater because it keeps them in the dark. That way they get used to the dark, and when they go back to their mother they don't notice the change. It's good to line the bottom of the box with newspaper for warmth, and a little shredded newspaper over that makes it like a nest. If your house is cool, you need a little more warmth. You can put the box, with the sweater over, under a light or near a heater.

Chicks don't need to eat until they're 36 hours old. But they'll perk up faster in the house with food and water, especially water. Fill a little saucer or jar lid with water. Set a stone in it, so they don't tip it over or get their feet too wet. If they don't drink, you just dip their beaks in. Have the water a little warm, or at least room temperature. Dip their beaks, and if one sees the other drink, they learn from that. It's very important they get water when they're really young. They drink a lot.

I put the feed in a separate dish. It's not good to let the feed get wet, so I put feed at one end of the box and water at the other.

Because if the feed does get wet, and they track that around, it gives them dysentery. If they get dysentery it's very hard to cure. The wet food sticks onto their feet and then they walk through it again and it gets contaminated.

For food, give them baby chick starter mash if you have it. And there's baby small cracked grain, chick scratch it's called. If you don't have those on hand, give them rolled oats or Quaker oats and chopped hard-boiled egg. Add a few bread crumbs with it, or a little cereal that's dry. Mix it in, sort of dry and crumbly. They eat that like everything. Then you can dry and crush the egg shells from the other hens to provide calcium, even for those little chickens. They eat that in a big way.

Sometimes you keep chicks in the house a few days. Even after that you can still give them back to the mother. It just depends how long the rest of the eggs take to hatch. They usually hatch in about 1½ to 2 days; but if another hen has laid an egg or two in the mother's nest, then those eggs take a little longer hatching.

BUYING EXTRA CHICKS

Most times the mother hen finishes a job in 2 days. Her chicks are all hatched out.

Then if you like you can buy more chicks to put under her. She can cover up to 6 more chicks than she can hatch, because her wing span spreads out. That's an economical way to build up your flock quicker than just letting her sit on those 10 to 14 chicks. A bantam hen can cover as many as 18 to 20 chicks. One lady did order 25 baby chicks, and she kept her bantam hen setting on some old eggs until the chicks arrived by mail. She put the whole 25 under her hen, and raised every one.

That was a large size bantam, maybe crossed with New Hampshire. If you cross a big New Hampshire rooster and a bantam hen, you'll get a bigger size bantam hen that still has the bantam instincts. And she's big enough to cover a lot of chicks.

If you buy extra chicks to put under the hen, any kind will do. But they should be the same age as her own chicks. Then the hen will take them well. The best age to get them is 2 or 3 days old.

MOVING THE HEN TO A "BEDROOM"

As soon as all the chicks are hatched, you should move the hen and her chicks to a private sheltered place for protection. A little coop about 4 feet by 4 feet will do well. Have a wooden box in there for shelter, and close the coop up at night. If you have to use the big chicken house, fix a little private cage for the hen and babies, so the other chickens won't bother them. After about 10 days, they can be with the others more.

PUTTING CHICKS BACK UNDER THE HEN

You may have bought extra chicks. Also you've kept the little first-hatched chicks in your house. Now you need to give them to the hen as soon as she's hatched out all her eggs. In either case, the procedure is the same. You have to sneak them under her when it's dark at night. Make her "bedroom" dark, so that until about 10 or 11 o'clock the next morning, when she starts feeding the new chicks, they're getting used to all the other chickens' voices and are acquainted with her voice. Then she doesn't pick at them. But if you get the chicks when they're older and you hope she'll take them — then maybe one won't obey her, and maybe 2 or 3 won't obey her. She picks and picks and picks at them. It's easier with

1 to 2 day old chicks. But some hens are more tolerant than others too. So be very careful to slip the chicks under her when it's dark. If they're starting to eat already you have to bring them in your house first and let them fill up with water and food before you put them under the hen. Then they can last until about 10 or 11 the next day.

The hen only has to pick at a chick a few times. Then it won't go near her any more. It acts like an outcast. They have a wonderful memory. If they've ever had another mother, they remember their mother's voice. If the hen pecks at them and they feel outcast, they go with their heads in a corner. She won't peck at their tail parts, she pecks at their heads. She pecks and pecks until they won't go near her any more. Then the only thing to do is try to get them under another mother. I had an outcast chick once and I gave it to a boy and told him how to put it under a hen. But he didn't do what I told him, and next day he brought it back and left it on the front porch in a bag. So I sent it to a friend the next day and told her what to do, and she did it and the chicken survived.

CHANGING MOTHERS

Sometimes a hen will only have about 3 or 4 chicks, and you'll say, well, I'm not going to let her waste her time on these 3 or 4 chicks. I'll give them all to the other one. Well, you can't do that, unless you make the room dark and give the chicks a good chance to get acquainted with that new mother's voice. And put the first mother far enough away where the chicks can't hear her voice. Because if they can hear, they would remember. It's remarkable the memory that they have.

RAISING MOTHERLESS CHICKS

Chicks can be hatched in an incubator. It's a little tricky, but you can do it. You're much better off with a broody hen, though, even if you have to borrow one. Then the hen does all the work instead of you.

After chicks are hatched, they can also be raised without a mother hen if necessary. You can keep them in a box with a warm light over it until they have feathers, about 6 weeks. But they're kind of nervous that way, like orphans. If they're raised with a mother, she teaches them how to take precautions when the cat comes around. She keeps them alerted. But without a mother they have no one to teach them. It works much better with a mother hen. You know where you're at then.

"the first lesson she gives them really is an obedience lesson..."

III.

HEN AND CHICKS

After her chicks are all safely hatched, the hen's quite contented. She's so proud of her family, she acts like what a wonderful thing it is to be a mother. She Keeps the chicks under her for a day or so, before they want to start eating. She prefers dry food, like baby chick starter mash and chick scratch, for her babies. She also finds them bugs. She begins to pick up the food and drop it down in front of them. She calls them. She says, "brrp, brrp, brrp, ... brrp, brrp, brrp," in a deep tone, and they all come around her. Then she picks up a little food and drops it down. If the food is not good, she says, "Krrrk," in a low tone, meaning, "Don't touch it!" And if the food is all right, she says, "drrrp, drrrp, drrrp," in a higher pitch, and they all come and eat.

Afer they've eaten and drunk, she calls them under her again, until about 2 or 3 hours and they're hungry again. Then they come out and eat more and drink more. They drink a lot of water. I put rocks in a saucer or shallow container of water. Then they don't fall in and drown with it being too deep. Also the chicks can stand on the rocks and drink without getting their feet too wet. It's always better if they don't walk in the water and then walk in their food, because they'll get the food wet and that never agrees with them. It gives them dysentery. It sticks on their feet and bottoms. Then it builds up, and you need to scrape it off. Then put vaseline on the chicks and watch them.

The hen and chicks should stay inside the chicken house, or their own little coop, for about 10 days. Then the chicks begin to get so their legs are stronger. They're ready for the mother hen to take them outside. She scratches in the soil for them and finds little bugs and little worms, and she calls them all together. You know how they say, "Children and chickens are always a-pickin'." So, the first lesson she gives them really is an obedience lesson, I think. She gives a reward for coming. She never calls them unless she has something for them.

The hen never leaves her chicks for any length of time to get cold. Soon as they commence to "peep peep" like they're unhappy, she calls them under her. She spreads out her wings and they can all get under her. She spreads her wings real wide. The feathers of her wings are almost like little pages where they can get the air under. They can peek out from her wings, under the feathers, and then get back under her again. When the weather is warmer, the chicks will climb up on the hen's back and ride piggyback. They have so much confidence in her.

In the morning, the hen gets her chicks up real early for food and water. She also gets them to bed real early at night. When she sees the sun going down, she starts to think they ought to go to bed. So she calls them all together and lets them eat their supper before bedtime.

Sometimes they get to bed, but one chick gets lost under a building and can't find its way out. The hen mothers the others, and she sits and keeps talking to that chick, calling that one in. If it can't find the way in, then finally she leaves the others to try and get it. By that time the chick is real cold, and she sits on it to get it warm. Now if the others don't go out and find her that's very bad, because the others get chilled too, but there's enough of them to get in a little corner by themselves and stay warm and huddle up.

If I happen to be home and I'm out there then, I take the mother hen out and I let her call the lost chick. Then once she gets it out of there, I can pick her up along with the stray chick, and put her with the group of them again. She calls her chick in a special way. She says, "BK, bK, bK, bK, bK.... BK, bK, bK, bK, bK ..." in a high tone, and then she sort of purrs. She wants to mother the chick. They all Know what she's telling them.

When there is danger of hawks or owls or cats or anything around, the hen gives out a warning signal. She'll get up somewhere and make a squawking noise. Then she says, "crrrK, crrrK, crrrK," in a low-pitched voice. The chicks all get under her, or all stand at attention. The hen stands guard until the enemy is gone. Then she gives the "all clear" signal, calls them out like "Bllpp, bllpp, bllpp," in a high-pitched voice, and they all come out and eat.

The hen trains her chicks every day, until they begin to get feathers and she's feeling very well herself. Then they can scratch for themselves. And if she gets a nice soft place where the dust is dry, she shows them how to take dust baths to Keep their skin clean. They close their eyes for a dust bath, to Keep the sand and dust out. They dig a little fresh dirt around the yard somewhere where the dust hasn't been disturbed, so there won't be any fleas in it. The hen scratches with her feet and beak until she makes the soil soft and loose. Then she goes on one side and rolls on her back, and on the other side until she gets her feathers full of dust. She shakes all the dust out. Then she feels clean.

The hen watches over her chicks until they're 2 months old. Then they can go on their own. If the hen is a bantam hen, she's ready now to start another family.

"They like the secure feeling of a roof right close over their backs."

IV.

ROOM AND BOARD FOR CHICKENS

FEED

I usually give the young chickens grain until they're about $3\frac{1}{2}$ to 4 months old. Then I start giving them egg maker for their developing days. They start to lay when they're $4\frac{1}{2}$ to 5 months old. You should have separate areas for those eating more grain. They all need to have oyster shells too. The calcium in the oyster shells makes their egg shells hard.

What they specially like is green food. We feed them grain night and morning, and they help themselves to whatever nature grows for them. They go for green food of any kind, like clover and alfalfa, and all the newer little grasses coming up. Soon as they're let out in the morning, first thing they do is hunt out to get green things. Which shows you that they know what's good for them. They love dandelion greens and small chickweed. My mother used to chop up nettles for them in England. I don't know how she did it. They've got such good eyes to spot green things. I think that's why they never have to wear glasses!

In the bug line, they like beetles and earwigs and small worms. Earwigs especially. They help keep your garden free of bugs. I had an apple tree with a crotch in it, and at night I used to take an old newspaper and roll it up and put it in that crotch. In the morning it'd be full of earwigs because the earwigs like to crawl in the newspaper at night. I'd take that out and shake it, and every time the chickens saw I was about to take that paper out, they'd come running right away.

They need plenty of drinking water. You can buy tablets to purify the drinking water. Use them once a month. If a person has milk for them, they love that too. Sour milk is best.

Funny how they know what they want to eat. I wonder if it's smell. They know when it's a sour apple or a sweet apple. If it's a sour apple they won't touch it. But if it's one of the Gravensteins coming down, they go for that in a big way. They keep waiting around under my tree. I made up my mind last fall they weren't going to have those apples. Then every time I'd go out and the wind had been blowing, why they'd have an apple half eaten before I got there. Just make a hole in the middle and just keep going.

Chickens eat many kinds of food left over from people's meals. They like bread scraps and most kinds of fruit and vegetables, just like birds. Old lettuce is very popular. They sure like pears, the sugar in pears. You know when a thing is safe to eat. Chickens sure won't eat it unless it's safe. I used to try out mushrooms on chickens. If it isn't a mushroom, they won't eat it. They won't eat a toadstool. They won't eat onion, green pepper, or cabbage, or citrus fruits. Also, I dry my eggshells so they're brittle, and then break them up for the chickens, to build up their calcium. But they still need oyster shells for the calcium.

Chickens need to eat little tiny rocks and pebbles. That grinds up the food that's inside them. When you open up a chicken, you always find all those little rocks in the gizzard. I've often seen them scratch in the manure on the perch. I wondered why. Well, when they don't get enough little rocks, they eat them over again. They have to have them.

HOUSING

The best kind of chicken yard is one with some gravelly places, for getting little rocks. A good chicken yard has part shade and part sun. The fence should be 6 feet tall. I like to have a place under the chicken house with soft soil for dust baths. Chickens like to take a dust bath every day. Get the dirt onto their feathers and then shake it all out. Then they feel pretty good. They love to lie and bask in the sun like people would.

A chicken house need not be large. For 12 chickens a 6 by 8 foot house is adequate. Nesting boxes and a roost are required furniture. They need feed and water in their house. It's important to have several windows to open for fresh air in summer. Chickens need fresh air. For a larger chicken house, have the south or southwest side open, but covered with wire. I usually have a low little sliding door to close easily at night when all the chickens are inside, safe from raccoons and foxes and stray dogs.

Your chicken house needs a wood floor, covered with dry straw or shavings. Nest boxes should be at least a foot square. And I hang a strip of burlap all the way along their nest boxes. Split it up, so it's like little panel curtains. That way each hen can lay her egg in her box and duck out her little curtain. They like privacy when they lay.

Perches are long rails for chickens to sit on and sleep huddled up at night. They curl their toes around them. They like their perches up high next to the ceiling. They like the secure feeling of a roof right close over their backs. Your bantams will roost in trees at night if you let them. But I don't let mine, because owls and things may get them.

MAINTENANCE

It's good to lime their whole house out once a year. You mix a paste of lime with water and brush or spray it on. Also lime on the drop boards. Nesting boxes should be cleaned once a year. And turn over old soil in the chicken yard sometimes, for better scratching and dust baths.

MEDICAL CARE

MITES. When chickens go up on the perch at night, eventually mites will develop where they've been sitting on the perch. You have to keep the perches clean with either used crank-case oil or stove oil, or some disinfectant to keep the mites down. They crawl up the chickens' legs and get on the chickens' bodies. They feed at night on the chickens. Then the chickens are tormented and they don't lay as well. They like to have a good night's sleep like we do.

You can get rid of the mites by using a small piece of rag or an old paint brush. Wad it up and tie it on a stick, and dip it in used crankcase or stove oil. Smear it all the way along on the places where they sit to sleep on the perches. Underneath the perches especially, because that's where the mites sleep all day.

Another kind of mite gets on the chickens' legs, and their legs get scaly. Rubbing carbolated vaseline on their legs at night should help this problem.

FLEAS. Fleas sleep all day, but they'll start feeding on the chickens any time after 4:00 in the afternoon. They especially develop where hens have been nesting for a while.

One reason bantam hens prefer to nest outside is that it's clean and there aren't any fleas. Especially early in the spring, you'll have fleas in the chicken house where they laid their eggs all winter. A way for hens to get rid of their fleas for a while is to take a dust bath.

One time I had a mother hen troubled with fleas, after she'd hatched her babies in the chicken house. I thought, well, I'd dust the mother hen with flea powder. I ruffled up her feathers and dusted her with flea powder. Then the chicks all went under her and they all died. They all got dosed with the flea powder. So dust baths are much better for controlling fleas.

PECKS. Some chickens will bite each other. And the different ages will peck at each other. The only way you can do if they persist is to take fingernail clippers and clip a little off their beak, just that little hook part. If they pick then, they won't pick quite so sharp, but they can still eat.

Sometimes chickens peck at each other so much they start bleeding. That starts what we call cannibalism. Then you have to put iodine on the bleeding place. When the other chickens smell the iodine, they won't peck at it.

SUPERVISION

The rooster is a sort of supervisor. He calls the chickens to eat sometimes. He looks after them and warns them of danger around. He goes to bed when they do. Sometimes he'll let them go first. Sometimes he'll go first and call them in there.

"If 2 roosters get fighting hard..."

The rooster can also act as referee. When 2 hens start fighting, the rooster will go between them and settle the fight. Then if they're not satisfied and start fighting again, he'll go back over and talk to them again, to say, "shame on you," and then they quit.

If 2 roosters get fighting hard, the only way I can do is to take a board and slap one of them in the face. I did that one time and the rooster never did know what hit him, so he gave up fighting.

I had one too that pecked at my legs, so I swatted him with a piece of board. He watches out for me now.

"If you feed them at night, they lay early in the morning."

V.

EGGS

After the chicks are 2 months old, when you discover which are roosters and which are pullets, you can separate them and fatten the roosters up for the table. The pullets, or young hens, you keep until they lay. Pullets that hatched out in February or March should start to lay in late August or September. Soon the older hens will all be molting, which means losing their feathers. They don't lay when they're molting. So it's good to have the young pullets laying.

Of course you always need to keep a rooster with the hens, if you want to have fertile eggs. You need one rooster to "service" 15 hens. Hens will lay eggs without a rooster, but the eggs won't be fertile, they won't hatch. And it's perfectly all right to eat fertile eggs, or non-fertile eggs. They taste the same. Sometimes people wonder about that.

LAYING

Chickens go to bed in the afternoon, just as soon as the sun goes down or while they're still very warm. They sleep warm then and stay warm. They don't like to stay around and shiver and then try to go to bed and get warm on the perch.

They'll go to bed soon after you've fed them, so in wintertime I feed mine anytime after 3:00; summertime, not till around 5:00 or so. They always fill up on all the food they can eat and drink before they go to bed. That's the nature of them. Then they hustle off to bed. Once they go up to sleep on the perch at night, unless you go in and put some more food down or something, they don't get up again.

The main thing is to keep them happy. Have dry straw or litter in the place where they sleep, so that when they get up in the morning they can scratch around in that dry stuff.

I put feed and water in for the chickens at night, so when they get up in the morning they can start eating and drinking right away, at daybreak. If you feed them at night, they lay early in the morning. But if you don't have feed in for them, they'll have to eat later. That way they lay their eggs later in the day, and they'll skip sometimes 2 whole days of laying a week because they haven't been fed early.

If you have an electric light or floodlight that comes on at 5:00 or 6:00 in the morning, take a flashlight at night and sprinkle around a little grain in the litter. Then when the chickens get up in the morning, they'll scratch for that grain. And as long as they're finding something, they'll keep on. That makes them work, and then you get early eggs. If you get eggs early in the morning, then usually they lay well all week. But if you're late in letting them eat, then they're going to be late in laying an egg for you too. It doesn't take them long. If they decide to lay an egg, it doesn't take them long to get in there and lay an egg, and then get out.

The eggs inside the chicken hang like a bunch of grapes. First come larger ones, and then they taper off to real small ones not much bigger than a grain of sand. I think when the chickens have to go to bed early at night, the work is going on inside of them, putting the shell on the egg, ready for the next morning's laying. Sometimes an egg's shell is soft when first laid, but hardens up quickly afterward, sort of like jello setting.

I don't think brown eggs are any healthier than white eggs. White chickens and brown chickens all eat the same Kind of food. People just like those lovely big round brown eggs. I do think one or two white eggs in the box bring out the value of the brown eggs. I think they look real pretty.

TESTING FOR LAYING

You can test a hen to see if she's laying. There are two bones either side of a hen's rectum. You see, a hen only has one vent for everything. If you can fit two of your folded Knuckles between those two bones, then the hen is laying. Otherwise not. The technique takes a bit of practice, but it's worth it. The longest it pays to Keep a hen is about 2½ to 3 years.

MOLTING

A hen may lay every day until the time to change feathers, to make new feathers. This is called molting. Sometime in the fall, she starts shedding the old feathers and making the new feathers. She takes 6 weeks to 2 months off from laying. And the more green stuff you can give her and get her to eat, and Keep her happy, the quicker she gets over her molt. In molting, the new feathers come along just like children's teeth and push the old feathers out. Hens really like a little cod liver oil when they're molting. Some of the feed is supposed to have some in.

KEEPING FERTILE EGGS FOR SETTING

Don't refrigerate or wash them, but Keep them in a cool room. They'll Keep for 4 to 5 weeks. It's best to turn them over every day.

TESTING EGGS FOR FERTILITY OR DEVELOPMENT

When you open an egg to wonder if it's fertile, there's a little cloudy spot on each side of the yolk. That means the egg is fertile. If there's only one cloudy spot on one side of the yolk, that means the egg is not fertile.

When a hen has been setting on eggs and you're not sure whether they're fertile and developing, take half a pail of warm water and put all the eggs in, on about the 15th or 16th day. Watch them closely. If they're not fertile, they sink to the bottom. If they have chicks inside, they float on top. But close to hatching time, they kick around in the water. That way you know they're developing, and you can expect chicks in 3 to 4 days. Only keep the eggs in water 3 minutes, no more; then hurry them back to the hen and her body heat will dry them.

WARMING CHILLED EGGS

If her eggs get chilled, when the hen is shut out of her nest somehow, she's very worried. She'll try every way to get in. But if she can't, and her eggs get real cold, then maybe you can save the eggs by warming them. If it hasn't been all night, try putting them in half a bucket of good warm water for a few minutes, and then back in the nest. The hen will set on them, and maybe the chill won't hurt the eggs. It might just set them back one day from hatching.

SENSITIVITY OF CHICKS IN EGGS

Chicks are sensitive, even in the shell. I've known the shock of a severe thunderstorm to kill all the chicks in a hen's nest of eggs! The hen will know it. Then she won't go back on that nest again. So you see how much even unhatched chicks need to feel secure.

"With bantams, the whole idea is to become mothers."

VI.

VIRTUES OF THE BANTAM HEN

SETTING AGAIN

When early spring chicks are about 2 months old, they can find their own food and get along without their mother. Then the mother bantam hen gets broody again. She's ready to start another family. She finds a new place for a nest. She pecks at her 2-month-old chicks if they come near her, and abandons them because she's taught them to go ahead. She makes them go off on their own. She doesn't want them to see where she goes and starts her new nest. Sometimes they're inquisitive and they go and look. She discourages them and scolds them. Then they decide "Mother's left us," and that's that. They go on about their business.

The bantam hen begins laying eggs again in her new nest. You see, a hen never lays eggs while she's busy mothering her chicks. She starts up again when the chicks go on their own. From then on it's the same story again. The hen lays 10 to 14 eggs, then sets on her new nest. After 21 days of setting on eggs, the mother hen brings out another family. Bantams can have as many as 4 families in a year.

A real broody bantam hen is so determined to be a mother, she'll stay on her nest if there's just one egg, and sometimes even if there's nothing!

Maybe you've been taking her eggs away every day. Well, then, if you see she's so persistent, you can put some more eggs back. She'll go right ahead and set.

MOVING A BROODY HEN

If you need a broody hen, maybe you can borrow one. Or sometimes you need to move your own broody hen to a particular spot. Be sure to box her in and keep her "bedroom" dark. Keep warm eggs under her too. There's a good chance she'll stay broody.

HATCHING OTHER KINDS OF EGGS

A bantam hen will set on other kinds of eggs too. Duck eggs, guinea eggs, or turkey eggs take 28 days to hatch. Goose eggs take 30 to 35 days. A good-sized bantam can hatch out about 5 duck eggs, or 2 or 3 goose eggs, or 11 guinea eggs, or 9 turkey eggs. She'll even turn the large goose eggs over every day! Then after the eggs hatch, she'll still bring up the babies, whatever kind they are.

EATING FOR "BROODINESS"

A bantam hen just naturally has a good mother instinct. She eats food that keeps her body warm, so she can mother more chicks. She eats more grain and bugs than regular hens, and that keeps her body warmer so she can produce more families. Just as soon as the weather gets nice, bantam hens begin to want to set. With bantams, the whole idea is to become mothers.

Whereas the other hens don't seem to mind. They don't want to become mothers as much, most of them. The regular hens will eat more egg-maker feed that doesn't build up heat in their bodies. They just go and lay their eggs and go on along. Still, if they were all mothers, there wouldn't be very many eggs to sell.

I found out that so long as hens eat the egg mash, they don't become broody. They fill up on it, and I try to make them fill up on it, because I want more eggs. But the bantam eggs don't sell very well anyway because they're so small. That's another reason people use bantams for mothers.

For instance, suppose you had a flock of 25 bantam hens. At least 23 of them would get broody in spring and want to be mothers. With a flock of 25 regular hens, you might get only 3 or 4 to be broody. We used to say, "The egg-maker feed really takes the broodiness right out of them." Before the egg-maker came out, the non-bantam hens used to want to be mothers also. But with egg maker available now, the bantams naturally still try hard to eat more grain. They'll go to no end of trouble to find grain. They'll dig seeds out of little weeds and such that come along so they can become mothers.

If you get a real good bantam that you know has good habits, you can expect some of her children to inherit the same good habits. That's good to know, because the longest you should keep a hen is $2\frac{1}{2}$ to 3 years.

People used to say if you had a rooster with the hens they'd become broody whether or not they were bantams, but that doesn't work.

FIGHTING SPIRIT

Another reason bantams make good mothers is their spirit. One thing I saw once. We had a cow grazing in the orchard near a bantam hen with chicks. The cow kept going closer and closer to the hen, and she thought the cow was going to step on her chicks. She was so upset about that. She started pecking at the cow's nose, fluffing her wings out and going after it, as if saying, "Don't you dare try to touch my babies!" The cow finally hauled off. It made me laugh to see a bantam try to fight a cow. That little thing fluffing her feathers out!

"Don't you dare
try to touch my babies!"